# YOUR KNOWLEDGE HAS VALUE

AF136293

- We will publish your bachelor's and master's thesis, essays and papers

- Your own eBook and book - sold worldwide in all relevant shops

- Earn money with each sale

Upload your text at www.GRIN.com
and publish for free

# Outlier Analysis. A Study of Different Techniques

Priyabrata Mishra
Soubhik Chakraborty

**Bibliographic information published by the German National Library:**

The German National Library lists this publication in the National Bibliography; detailed bibliographic data are available on the Internet at http://dnb.dnb.de.

ISBN: 9783346702463
This book is also available as an ebook.

© GRIN Publishing GmbH
Nymphenburger Straße 86
80636 München

Print and binding: Books on Demand GmbH, Norderstedt, Germany
Printed on acid-free paper from responsible sources.

The present work has been carefully prepared. Nevertheless, authors and publishers do not incur liability for the correctness of information, notes, links and advice as well as any printing errors.

GRIN web shop: https://www.grin.com/document/1254838

# A STUDY OF DIFFERENT OUTLIER ANALYSIS TECHNIQUES

BY

## PRIYABRATA MISHRA
## SOUBHIK CHAKRABORTY


## DEPARTMENT OF MATHEMATICS
## BIRLA INSTITUTE OF TECHNOLOGY
## MESRA, RANCHI-835215, JHARKHAND
## INDIA

# PREFACE

Data mining is perhaps one of the most intriguing fields. The scope of data collection and analytics has risen tremendously as data is digitized and systems are networked and integrated. Most systems nowadays create non-stationary data of enormous size, volume, occurrence speed, and rapid change. This makes large-scale data analytics difficult. In any application that involve data, outlier detection is critical. In the data mining and statistics literature, outliers are sometimes known as abnormalities, discordants, deviants, or anomalies. The data in most applications are generated by one or more generating processes, which may reflect system activity or observations about entities. Outliers are created when the generating process behaves in an unusual way. As a result, an outlier frequently provides useful information regarding anomalous system and entity features that influence the data creation process. Recognizing uncommon traits can lead to useful application-specific insights.

This monograph explains what an outlier is and how it can be used in a variety of industries in the first chapter of the report. This chapter also goes over the various types of outliers. Outlier analysis is an important part of research or industry that involves a large amount of data, as described in Chapter 2; it also describes how outliers are related to different data models. Chapter 3 covers Univariate Outlier Detection and methods for completing this task. Multivariate Outlier Detection techniques such as Mahalanobis distance and isolation forest are covered in Chapter 4. Finally, in Chapter 5, the Python programming language has been used to analyse and detect existing outliers in a public dataset. We hope this monograph would be useful to students and practitioners of statistics and other fields involving numerical data analytics.

**Priyabrata Mishra**

**Soubhik Chakraborty**

# ACKNOWLEDGEMENT

I express my deep regards to my project supervisor **Dr. Soubhik Chakraborty**, Professor and ex-Head, Department of Mathematics, Birla Institute of Technology, Mesra, Ranchi under whose guidance I was able to learn and apply the concepts presented in this project. His consistent supervision, constant inspiration and invaluable guidance have been of immense help in carrying out this project work with success.

I am very grateful to **Dr. S.K. Jain,** Professor and Head, Department of Mathematics, Birla Institute of Technology, Mesra, Ranchi for extending all the facilities, and giving valuable suggestions at all times for pursuing this course.

I am also thankful to all the faculties of the department and other staff for their help and suggestions during the project.

Priyabrata Mishra

# Table of Contents

# LIST OF FIGURES

# CHAPTER 1: WHAT IS AN OUTLIER & ITS TYPES

Outliers are observations or measurements that are unusually tiny or huge in comparison to the vast majority of observations. In the data mining and statistics literature, outliers are sometimes known as abnormalities, discordant, deviants, or anomalies(von Eye and Schuster, 1998). Outliers are created when the generating process behaves in an abnormal way. As a result, an outlier frequently provides useful information regarding anomalous system and entity features that influence the data creation process. The issue is that a few outliers can occasionally cause the group results to be skewed (by altering the mean performance, by increasing variability, etc.).

Examples of Outliers causing problems:

- Various forms of data concerning operating system calls, network traffic, and other user actions are collected in many computer systems. Because of malicious activities, this data may indicate strange behavior. Intrusion detection is the process of detecting such activities.

- Credit card fraud has become more common as the ease with which sensitive information like a credit card number can be hacked has increased. Unauthorized credit card use can manifest itself in a variety of ways, such as shopping sprees at certain areas or extremely big transactions. Outliers in credit-card transaction data can be detected using such patterns.(von Eye and Schuster, 1998)

- Many entries relating to patient diseases, treatments, and lab findings can be found in patient medical records. These usually involve a variety of data kinds and generate a big amount of data. These databases can give critical information for clinical decision-making and hospital management. Medical databases contain several unique characteristics that are rarely found in non-medical databases. Outlier detection techniques can be used in this context to detect anomalous trends in health records (for example, data quality issues), resulting in better data and information in the decision-making process. (Gaspar *et al.*, 2011)

- Fig 1 depicts the outlier detection flowchart. The yield of products will be affected by various parameters and machines. As a result, when the outlier detection module receives the selected data, it divides the log files into several files based on the recipe number and then the tool number. The outlier detection module processes the separated files using the MapReduce technique to calculate means and standard deviations after obtaining them. The outlier detection module performs outlier detection after obtaining the means and standard deviations of each parameter.

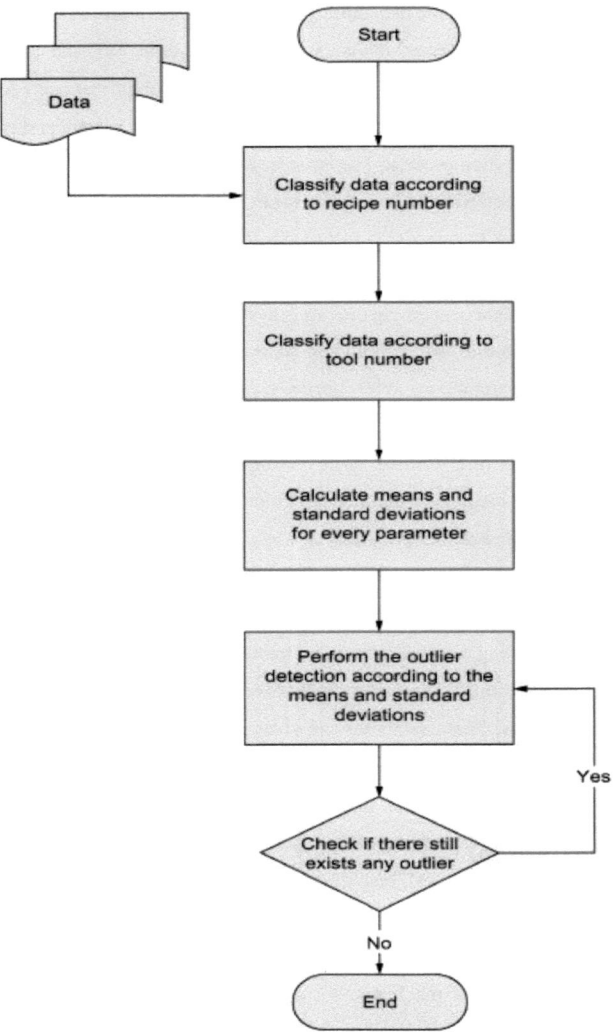

(Fig 1: Flow chat – Outlier Detection in IOT use case)

## Types of Outliers

- Type 1: Global Outliers
- Type 2: Contextual Outliers
- Type 3: Collective Outliers

# Global Outliers

Point Outliers is another name for them. Outliers of this kind are the most basic. A global outlier is a data point in a dataset that deviates significantly from all other data points. Outlier detection methods are mostly used to find global outliers.

In an Intrusion Detection System, for example, if a large number of packages are broadcast in a short period of time, this may be considered a global outlier, and we can conclude that the system has been hacked.

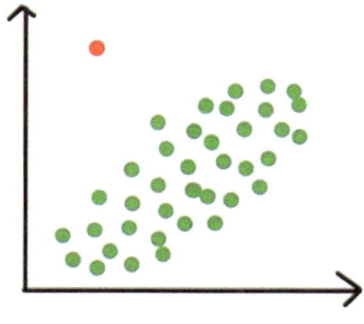

(Fig 2: Global Outliers)

In Fig 2, red data point is an outlier to the dataset.

# Contextual Outliers

Conditional Outliers are another name for them. If a data object in a dataset deviates significantly from the other data points because of a single context or situation. Due to one situation, a data point may be an outlier, yet under another environment, it may behave normally. In order to discover contextual outliers, a context must be included as part of the problem statement. Contextual outlier analysis gives users the ability to study outliers in diverse situations, which is useful in a variety of applications. Both environmental and behavioral attributes are used to determine the data point's qualities.

For example, in the context of a "winter season," a temperature reading of 40°C may act as an outlier, but in the context of a "summer season," it will behave as a normal data point.

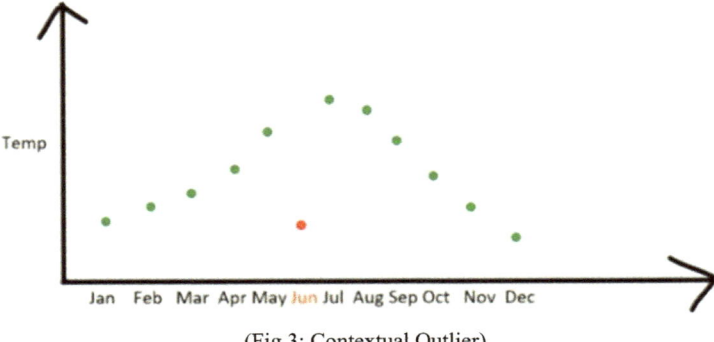

(Fig 3: Contextual Outlier)

In Fig. 3, it can be noticed that the low temperature in June is a Contextual Outlier, because the same value is not considered as an outlier for the month December.

## Collective Outlier

Collective outliers, as the name implies, occur when a group of data points in a dataset deviates significantly from the remainder of the dataset. Individual data objects may not be outliers in this case, but when viewed as a group, they may act as outliers. We may require background knowledge about the link between the data objects exhibiting outlier behavior in order to discover these types of outliers.

For example, a DOS (denial-of-service) package sent from one computer to another may be considered regular behavior in an Intrusion Detection System. However, if this occurs on numerous computers at the same time, it may be regarded abnormal behavior, and they might be classified as collective outliers as a whole.

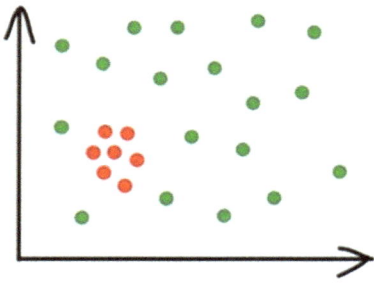

(Fig 4: Collective Outlier)

The red data points in Fig 4 as a whole are collective outliers.

# CHAPTER 2: OUTLIER DETECTION IMPORTANCE & ITS CONNECTION WITH DATA MODELS

## Importance of Outlier Detection

A distribution's extreme value might be lawful or illegitimate. Returning to the perfectly balanced coin, which lands on the 'heads' 100 times out of 100. It would be a mistake to leave such an observation out of a planned research because it is a genuine observation that should not be modified if the coin is correctly balanced. If, on the other hand, the coin seems to be balanced but is actually a rigged coin with a 0% probability of giving 'tails,' then leaving the data alone is the wrong approach to dealing with the outlier since it represents a value from a different distribution than the one of interest. Changing (e.g., excluding) the observation in the first circumstance results in a variance reduction that is insufficient since a value from the considered distribution is deleted. Leaving the data alone in the second case, on the other hand, means under-enlarging the variance because the observation is not from the distribution that supports the experiment. In both cases, a poor decision can alter the test's Type I error (alpha error, i.e., the probability that an incorrect hypothesis is not rejected) or Type II error (beta error, i.e., the chance that an incorrect hypothesis is not re jected). Making the correct choice has no bearing on the test's error rates.

However, there is a significant difference between Grubbs' time and today's time. The volume and speed with which data is created and processed is greater than it has ever been. Millions of social media posts, messages, transactions, and videos are created every second. As a result, outlier detection algorithms must be able to process data in near-real time. They need to show potential outliers as soon as possible. Because the information provided by outlier detection in Big Data is often time sensitive. The time-sensitive nature of outlier detection results is demonstrated by the examples we discussed earlier, such as credit card detection and malicious chatter.

As a result, it is not exaggerated to say that Big Data has revolutionized outlier detection. Simultaneously, Big Data has opened up a whole new world of possibilities for extracting (more) value from outlier detection techniques. Because the size of the data set grows larger, outlier detection may become more valuable. Consider the (famous) task of finding a needle in a haystack as an example. The more valuable an outlier detection algorithm becomes, the larger the haystack becomes.

## Connection of Outliers with Data Models

A model of the data's typical patterns is built by almost all outlier detection algorithms, and the deviations from these patterns are used to calculate an outlier score for a given data point. This data model could, for instance, be a generative model based on regression, proximity, or a Gaussian-mixture model. Different suppositions are made by each of these models regarding the "normal" behaviour of the data. Different suppositions are made by each of these models regarding the "normal" behaviour of the data. After that, the outlier score of a data point is calculated by assessing how well the data point and model fit together. The model may frequently be defined algorithmically. For instance, nearest neighbor-based algorithms for outlier detection model a data point's propensity for being an outlier in terms of the distribution of its k-nearest neighbour distance. The assumption in this situation is that outliers are spread out from the majority of the data.

The data model you choose is very important. The results could be subpar if the data model is chosen incorrectly. A fully generative model, like the Gaussian mixture model, for instance, might not perform well if the data does not match the model's generative assumptions or if there are not enough data points to learn the parameters of the model. Similar to this, if the underlying data is clustered arbitrarily, a model based on linear regression may not perform well. Because of the poor fit to the incorrect model assumptions in these situations, data points may be misreported as outliers. Unfortunately, learning the best model for a given data set for outlier detection is largely an unsupervised problem without examples of outliers. Outlier detection is different from many other supervised data mining problems in that it lacks labelled examples, which makes it more difficult to solve. As a result, in real-world situations, the analyst's understanding of the types of deviations pertinent to a given application frequently determines the model to be used. For instance, it would be reasonable to assume that an unusual deviation of the temperature attribute in a spatial locality is an indicator of abnormality in a spatial application measuring a behavioural attribute, such as the location-specific temperature. On the other hand, due to data sparsity in the case of high-dimensional data, even the definition of data locality may be unclear. As a result, only after carefully examining the pertinent modelling properties of that domain can an effective model be built for a given data domain.

The choice of a model involves many trade-offs; a highly complex model with an excessive number of parameters will probably overfit the data and also find a way to fit the outliers. A straightforward model that is built with a solid intuitive grasp of the data (and perhaps also a grasp of what the analyst is seeking) is likely to produce much better outcomes. On the other hand, a model that has been oversimplified and does a poor job of fitting the data is likely to label common patterns as outliers. Perhaps the most important step in outlier analysis is the initial choice of the data model. Throughout the book, the theme of data models' impact will be repeated with specific examples.

The outlier detection problem can be seen as a classification problem variation where the class label (either "normal" or "anomaly") is not present. As a result, one can "pretend" that the entire data set contains the normal class and create a (possibly noisy) model of the normal data because the normal examples vastly outnumber the anomalous examples. Outlier scores are those that deviate from the norm. Because a lot of the theory and techniques from classification generalise to outlier detection, the relationship between classification and outlier detection is crucial. Outlier detection methods are referred to as unsupervised, whereas classification methods are referred to as supervised, due to the unobserved nature of the labels (or outlier scores). When anomaly labels are present, the issue can be reduced to the imbalanced form of data classification.

A one-class analogue of the multi-class setting in classification may be thought of as the model of normal data for unsupervised outlier detection. However, from a modelling standpoint, the one-class setting can occasionally be much more nuanced because it is much simpler to tell apart examples of two classes than it is to determine whether a specific instance corresponds to examples of a single (normal) class. The accuracy of the model can be sharpened by learning the distinguishing traits between the two classes more readily when there are at least two classes available.

There is a natural division between explicit generalisation methods and instance-based learning methods in many types of predictive learning, including classification and recommendation. This dichotomy also applies to the unsupervised domain because outlier detection methods need to create a model of the normal data in order to make predictions. A training model is not created in advance when using instance-based methods. Instead, one computes the most pertinent (i.e., closest) instances of the training data for a given test instance and then makes predictions on the test instance based on these related instances. In the fields of classification and recommender systems, instance-based methods are also known as memory-based methods and lazy learners, respectively.

# CHAPTER 3: UNIVARIATE OUTLIER DETECTION

Earlier univariate techniques for outlier detection are based on the assumption of an underlying known distribution of the data that is intended to be equally and independently distributed (i.i.d.). Many discordance approaches for spotting univariate outliers also assume that the distribution parameters and type of anticipated outliers are known. In real-world data-mining applications, these assumptions are routinely broken.(Zhao, 2013)

A generating model, which allows a small number of observations to be randomly sampled from distributions GI,..., Gk, rather than the target distribution, which is often assumed to be a normal distribution N, is a central assumption in statistical-based methods for outlier detection (p, u2). The problem of identifying those observations that fall within a "outlier region" is then translated into the problem of identifying outliers.

Hence the following definition:
For any confidence coefficient $\alpha$, $0 < \alpha < 1$, the $\alpha$-outlier region of the N $(\mu, \sigma^2)$ distribution is defined by

$$out\ (\alpha, \mu, \sigma^2) = \{x: |x - \mu| > Z_{1-\alpha/2}\sigma\}$$

where $z_q$ is the q quintile of the N(0,1). Although the normal distribution has traditionally been the target distribution, this approach may easily be applied to any unimodal symmetric distribution with positive density function, including multivariate distributions.

Note that, the outlier definition doesn't specify which observations are polluted, i.e., those deriving from distributions $G_1,..., G_k$, but rather those that fall within the outlier region.

## Standard Deviation Method

This is the simplest method to detect univariate outliers. The method is defined as:

$$X_{mean} \pm 2SD$$
$$X_{mean} \pm 3SD$$

Where $X_{mean}$ is sample mean and SD is the standard deviation of the sample.

All the observations outside this range are considered as outliers.

For a random variable X with mean $\mu$ and variance $\sigma^2$ Chebyshev inequality says that for any $k > 0$

$$P[|X - \mu| \geq k\sigma] \leq \frac{1}{k^2}$$

$$P[|X - \mu| < k\sigma] \geq 1 - \frac{1}{k^2}, \qquad k > 0$$

We may use the inequality $[1-(1/k)^2]$ to figure out what percentage of our data will be within k standard deviations of the mean. (Seo and Gary M. Marsh, 2006)

**Code Implementation:**

```
import numpy

arr = [854, 636, 17, 74154, 52, 758, 658, 1000, 457, 968, 362, 109, 451, 687]

elements = numpy.array(arr)

mean = numpy.mean(elements, axis=0)
sd = numpy.std(elements, axis=0)

final_list = [x for x in arr if (x > mean - 2 * sd)]
final_list = [x for x in final_list if (x < mean + 2 * sd)]
print(final_list)
```

[854, 636, 17, 52, 758, 658, 1000, 457, 968, 362, 109, 451, 687]

(Fig 5: SD Method for outlier detection code)

Fig. 5 the code for Standard Deviation method for outlier detection. It can be noticed that 74154 is found to be an outlier for the above data. Hence it was not included in the final list of outcomes which is outlier free.

# Z-Score method

The Z-score, which is known as the standard score, is a way of expressing a data point's relationship to the mean and standard deviation of the group. Getting a Z-score is as simple as translating the data into a distribution with 0 mean and 1 standard deviation.

Z-scores are used to remove the effects of data location and size, allowing for direct comparison of heterogeneous datasets. After the data is centered and rescaled. According to the Z-score approach of outlier detection, anything that is very large or less than 0 should be considered an outlier. The threshold point is usually a Z-score which is in between 3 or -3. See Fig. 6.

```
import numpy as np

def outliers_z_score(data):
    threshold = 3

    mean_y = np.mean(data)
    stdev_y = np.std(data)
    z_scores = [(y - mean_y) / stdev_y for y in data]
    return np.where(np.abs(z_scores) > threshold)
```

(Fig 6: Z Score method code implementation)

The above function calculates the outliers by applying the discussed Z-score method. If we take a dataset; for example, taking the array used in the last section as input dataset the result can be shown below:-

```
data = [854, 636, 17, 74154, 52, 758, 658, 1000, 457, 968, 362, 109, 451, 687]
print("posiiton of the outlier: "+ str(outliers_z_score(data)))

posiiton of the outlier: (array([3], dtype=int64),)
```

(Fig 7: Z Score output)

Fig. 7 shows that the outlier is at the position 3 which refers to 74154.

To quantify central tendency and dispersion, the Z-score approach uses the mean and SD of a collection of data. This is a concern since outliers have a big influence on the mean and SD, and they aren't very robust. In fact, one of the most important reason to find and remove outliers from a dataset is to reduce skewness in the data.

## Modified Z-Score method

Another disadvantage of the Z-score approach is that it acts oddly in tiny datasets; in fact, if the dataset contains fewer than 12 items, the Z-score method will never find an outlier. This prompted the creation of a modified Z-score approach that does not have the same drawback.(Iglewicz and Hoaglin, 1993)

```
import numpy as np

def outliers_modified_z_score(data):
    threshold = 3.5

    median_y = np.median(data)
    median_absolute_deviation_y = np.median([np.abs(y - median_y) for y in data])
    modified_z_scores = [0.6745 * (y - median_y) / median_absolute_deviation_y
                         for y in data]
    return np.where(np.abs(modified_z_scores) > threshold)
```

```
data = [854, 636, 17, 74154, 52, 758, 658, 1000, 457, 968, 362, 109, 451, 687]
print("posiiton of the outlier: "+ str(outliers_modified_z_score(data)))

posiiton of the outlier: (array([3], dtype=int64),)
```

(Fig 8: Modified Z Score method)

Fig. 8 shows the code implementation of Modified Z Score which gives the same result as previous cased if we take the same dataset.

The modified Z-score technique also has the advantage of using the median and MAD rather than the mean and SD. The MAD and the median are both reliable measurements of central tendency and dispersion.

## Interquartile Range (IQR) Method

The IQR approach of outlier detection is another reliable method for categorizing outliers. Because this was back when people used to do calculations and charting manually by hand, the datasets were often tiny, and the focus was on deciphering only the story of the data. This strategy can be seen in action if you've ever seen a box-and-whisker plot.

The data is plotted using quartiles to show the form of the data (data is divided into 4 equal parts/group by these points). The boxes show the first and third quartiles, which correspond to the 25th and 75th percentiles, respectively. The 2nd quartile, or median, is shown by the line inside the box.

The interquartile range is the area between the first and third quartiles, and it is this range that lends this method of outlier detection is named. Any data point outside of 1.5 times the IQR below the first – or 1.5 times the IQR above the third – quartile was considered "outside" or "far out" by Tukey. In a typical box-and-whisker plot, the "whiskers" extend all the way to the last data point that isn't "outside."

```
[2]  import numpy as np

     def outliers_iqr(data):
         quartile_1, quartile_3 = np.percentile(data, [25, 75])
         iqr = quartile_3 - quartile_1
         lower_bound = quartile_1 - (iqr * 1.5)
         upper_bound = quartile_3 + (iqr * 1.5)
         return np.where((data > upper_bound) | (data < lower_bound))

[3]  data = [854, 636, 17, 74154, 52, 758, 658, 1000, 457, 968, 362, 109, 451, 687]
     print("position of outlier: "+ str(outliers_iqr(data)))

     position of outlier: (array([3]),)
```

(Fig 9: Interquartile range method)

Fig. 9 gives the code for the Interquartile range method.

The interquartile range approach, same as the modified Z-score method, has the advantage of utilizing a robust measure of dispersion.

# CHAPTER 4: MULTIVARIATE OUTLIER DETECTION

In today's world, more and more seen data is multi-dimensional, increasing the likelihood of uncommon observations. The issue is that a few outliers are usually enough to skew data results by changing mean performance, increasing variability, and so on. As a result, finding outliers is becoming increasingly important in a variety of scientific fields, including psychology, finance, and chemometrics.

## The Mahalanobis Distance

The Mahalanobis distance is used significantly in multivariate statistics for detecting outliers. It is the distance between an observation and the mean of a distribution in standard deviations. This metric may be used to detect outliers since outliers do not behave like normal data in at least one dimension. It works well with multivariate data because it calculates the distance between data points and the center using a covariance matrix of factors. Unlike the Euclidean distance, MD finds outliers based on the distribution pattern of data points.

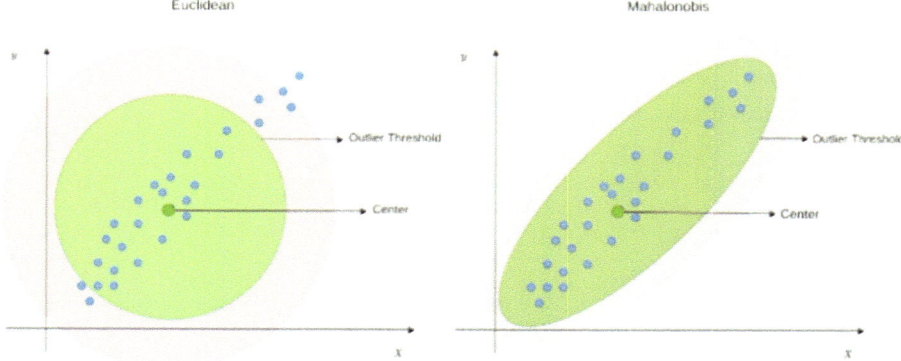

(Fig 10: Euclidean vs Mahalanobis distance ;reference: towardsdatascience)

As may be seen from fig. 10, data points are strewn in a certain direction. While Euclidean distance may classify some non-outlier points in such distributions as outliers, Mahalanobis distance can keep up.

The covariance matrix, which shows how variables vary together, is the major explanation for this disparity. Using covariance to calculate the distance between the centre and points in n-dimensional space allows for the discovery of real threshold borders based on variation. The covariance matrix had provided as C and the negative first power of it had taken, as can be observed from the distance formula of MD stated in the below formula. The Xpi vectors indicate observation coordinates in n-dimensional space. When looking for outliers, however, the distance between the centre and each observation should be determined rather than the distance between each point. The average value of each variable may be used to find the centre point.

The Mahalanobis distance formula is given by:

$$D^2 = \left( X_{p_1} - X_{p_2} \right)^T . \, C^{-1} . \left( X_{p_1} - X_{p_2} \right)$$

Where,

$D^2$ – Square of the Mahalanobis Distance

$X_{p1}$ – Vector of the observation

$X_{p2}$ – Vector of mean values of independent variables

$C^{-1}$ – Inverse covariance matrix of independent variables

Fig. 11 provides the code for Mahalanobis Distance as a classifier.

```python
class MahalanobisBinaryClassifier():
    def __init__(self, xtrain, ytrain):
        self.xtrain_pos = xtrain.loc[ytrain == 1, :]
        self.xtrain_neg = xtrain.loc[ytrain == 0, :]

    def predict_proba(self, xtest):
        pos_neg_dists = [(p,n) for p, n in zip(mahalanobis(xtest, self.xtrain_pos), mahalanobis(xtest, self.xtrain_neg))]
        return np.array([[(1-n/(p+n)), 1-p/(p+n)) for p,n in pos_neg_dists])

    def predict(self, xtest):
        return np.array([np.argmax(row) for row in self.predict_proba(xtest)])


clf = MahalanobisBinaryClassifier(xtrain, ytrain)
pred_probs = clf.predict_proba(xtest)
pred_class = clf.predict(xtest)
```

(Fig 11: Code implementation for Mahalanobis distance classifier)

We use Mahalanobis distances between a particular observation (row) and both the positive and negative datasets to predict the test dataset's class. The class is then allocated to the observation based on the group to which it belongs. In the above code snippet xtrain_pos is positive dataset and xtrain_neg is the negative dataset. AU-ROC curve, Accuracy score, precision and recall can be used as evaluation metrices for this classification.

## Outlier Detection using Isolation Forest

The primary concept of Isolation Forest varies from other common outlier detection algorithms in that it discovers abnormalities directly instead of analyzing typical data points. Isolation Forest, just like any tree ensemble method, is built on decision trees. In these trees, partitions are created by selecting a feature at random and then choosing an arbitrary split value between feature's min and max value.

Outliers are uncommon than normal observations, and they differ in terms of values and location in the feature space from the regular observations. As an outcome, they should be discovered with fewer splits nearer to the tree's root.

$$s(x, n) = 2^{-\frac{E(h(x))}{c(n)}}$$

where,

h(x)  –  Path length of observation x (Modelling, 2020)

c(n)  –  Average path length of unsuccessful search in a Binary search Tree (Modelling, 2020)

n  –  Number of external nodes (Modelling, 2020)

An anomaly score is given to every observation, which can be used to make subsequent decisions as follows,

- Score lies close to 1, it's a case of anomalies.
- Score much less than 0.5 is a case of normal observations.
- Score close to 0.5 is a case where the sample does not have clearly distinct anomalies.

# CHAPTER 5: OUTLIER DETECTION USING A DATASET

## Dataset Details

Many societies regard healthcare fraud to be a problem. Instead of being spent on medicine, geriatric care, or emergency department visits, health care funds are diverted to fraudulent operations by materialistic practitioners or patients. Healthcare fraud is a major contributor to increased healthcare expenditures, which is a result of rising healthcare prices.

Here a dataset created by The Centers for Medicare and Medicaid Services is used for the experimental work.

"The Centers for Medicare & Medicaid Services (CMS) has prepared a public data set, the Provider Utilization and Payment Data Physician and Other Supplier Public Use File (herein referred to as "Physician and Other Supplier PUF"), with information on services and procedures provided to Medicare beneficiaries by physicians and other healthcare professionals."(Centers for Medicare & Medicaid Services, 2020)

## Data Preprocessing

The raw data has to be preprocessed before being used to give better result. Preprocessing steps include removing unnecessary columns from dataset, cleaning data such as removing comma, imputing missing values, applying encoding to variables, applying standardization to the dataset etc.

**Removing Unnecessary Commas:** RemoveComma method was created for this purpose (fig. 12).

```
def RemoveComma(x):
    return x.replace(",","")
```

(Fig 12: RemoveComma method)

**Applying Encoding:** Binary encoding was applied to our data to convert categorical variables into numeric. BinaryEncoder method from the library category_encoders served this purpose.

**Standardization:** To make our data standard or normal, i.e., to make the mean 0 and variance 1, standardization is applied to each and every column of the dataset. We used StandardScaler function from sklearn.preprocessing library for this.

Attaching below the snapshots of the raw data and the data after all the preprocessing steps are done (fig. 13).

| index | National Provider Identifier | Last Name/Organization Name of the Provider | First Name of the Provider | Middle Initial of the Provider | Credentials of the Provider | Gender of the Provider | Entity Type of the Provider | Street Address 1 of the Provider |
|---|---|---|---|---|---|---|---|---|
| 0 | 8774979 | 1891106191 | UPADHYAYULA | SATYASREE | NaN | M.D. | F | I | 1402 S GRAND BLVD |
| 1 | 3354385 | 1346202256 | JONES | WENDY | P | M.D. | F | I | 2950 VILLAGE DR |
| 2 | 3001884 | 1306820956 | DUROCHER | RICHARD | W | DPM | M | I | 20 WASHINGTON AVE |
| 3 | 7594822 | 1770523540 | FULLARD | JASPER | NaN | MD | M | I | 5746 N BROADWAY ST |
| 4 | 746159 | 1073627758 | PERROTTI | ANTHONY | E | DO | M | I | 875 MILITARY TRL |

(Fig 13: Raw data)

Fig. 14 gives a snapshot of the data after all the preprocessing steps.

| | Average Medicare Allowed Amount | Average Submitted Charge Amount | Average Medicare Payment Amount | Average Medicare Standardized Amount | Credentials of the Provider_0 | Credentials of the Provider_1 | Credentials of the Provider_2 | Credentials of the Provider_3 | Credentials of the Provider_4 | Credentials of the Provider_5 |
|---|---|---|---|---|---|---|---|---|---|---|
| 0 | 0.385450 | -0.046433 | 0.400082 | 0.414299 | -0.097882 | -0.115257 | -0.129964 | -0.1553 | -0.193636 | -0.260789 |
| 1 | 0.086673 | 0.182805 | 0.207649 | 0.286359 | -0.097882 | -0.115257 | -0.129964 | -0.1553 | -0.193636 | -0.260789 |
| 2 | -0.041922 | -0.187794 | -0.064687 | -0.087154 | -0.097882 | -0.115257 | -0.129964 | -0.1553 | -0.193636 | -0.260789 |
| 3 | -0.380709 | -0.328957 | -0.370166 | -0.372921 | -0.097882 | -0.115257 | -0.129964 | -0.1553 | -0.193636 | -0.260789 |
| 4 | -0.291221 | -0.296019 | -0.289505 | -0.294800 | -0.097882 | -0.115257 | -0.129964 | -0.1553 | -0.193636 | -0.260789 |

(Fig 14: Data after all the preprocessing steps)

# Results

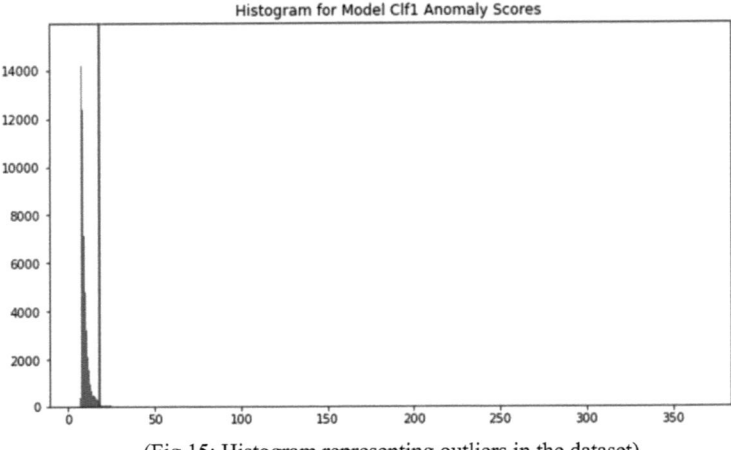

(Fig 15: Histogram representing outliers in the dataset)

Fig, 15 describes the threshold value beyond which data points are considered outliers. For our case, that value is 18. Principal Component Analysis was applied to reduce the dimension of the data, then PyOD library was used to fit the model which identified the outliers.

# CONCLUSION

The problem of outlier detection has applications in many fields where it is desirable to identify interesting and unusual events in the underlying generating process. The foundation of all outlier detection techniques is the development of a probabilistic, statistical, or algorithmic model that describes the normal data. The outliers are determined using the model's deviations. Designing straightforward, precise models that do not overfit the underlying data frequently requires strong domain expertise. When there are strong relationships between the various data points, the problem of outlier detection becomes particularly difficult. This is the case for time-series and network data, where the outliers are primarily defined by patterns in the relationships (whether temporal or structural) between the data points. There is definitely plenty of room for further investigation into outlier analysis, particularly in the structural and temporal domain.

# ABOUT THE AUTHORS

1. **Priyabrata Mishra** has completed his Integrated M.Sc in Mathematics & Computing from Department of Mathematics at Birla Institute of Technology, Mesra, Ranchi, Jharkhand, India. His research interests are applied and computational statistics. The present work is a part of his master's dissertation which he completed under the guidance of Prof. Soubhik Chakraborty.

**Priyabrata Mishra**

2. **Soubhik Chakraborty,** a PhD in Statistics, is currently a Professor and ex-Head in the department of Mathematics, Birla Institute of Technology, Mesra, Ranchi, Jharkhand. His research interests are algorithm analysis, music analysis and statistical computing. He has been guiding several research scholars in these areas leading to PhD and has published over 100 papers, 6 books and 7 research monographs. He is also an acknowledged reviewer associated with ACM, IEEE and AMS. He has received several awards in both teaching and research.

**Prof. (Dr.) Soubhik Chakraborty**

# REFERENCES

1) Centers for Medicare & Medicaid Services (2020) 'Medicare Fee-For-Service Provider Utilization & Payment Data Physician and Other Supplier Public Use File: A Methodological Overview', *Centers for Medicare & Medicaid Services*.

2) von Eye, A. and Schuster, C. (1998) *Outlier Analysis, Regression Analysis for Social Sciences*. doi: 10.1016/b978-012724955-1/50180-7.

3) Gaspar, J. *et al.* (2011) 'A systematic review of outliers detection techniques in medical data: Preliminary study', *HEALTHINF 2011 - Proceedings of the International Conference on Health Informatics*, (June 2014), pp. 575–582. doi: 10.5220/0003168705750582.

4) Iglewicz, B. and Hoaglin, D. C. (1993) *Volume 16: How to Detect and Handle Outliers", The ASQC Basic References in Quality Control: Statistical Techniques, Technometrics*.

5) Modelling, A. (2020) 'Fraud Detection in Financial Businesses Using Data Mining Approaches', (January).

6) Seo, S. and Gary M. Marsh, P. D. (2006) 'A review and comparison of methods for detecting outliersin univariate data sets', *Department of Biostatistics, Graduate School of Public Health*, pp. 1–53. Available at: http://d-scholarship.pitt.edu/7948/.

7) Zhao, Y. (2013) 'Outlier Detection', *R and Data Mining*, (1980), pp. 63–73. doi: 10.1016/b978-0-12-396963-7.00007-6.

# YOUR KNOWLEDGE HAS VALUE

- We will publish your bachelor's and master's thesis, essays and papers

- Your own eBook and book - sold worldwide in all relevant shops

- Earn money with each sale

## Upload your text at www.GRIN.com and publish for free